STO

FRIENDS
OF ACPL

104

MICROHABITATS

Life in a
GARBAGE
DUMP

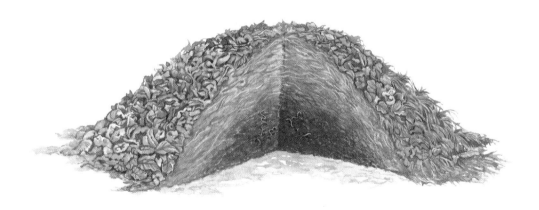

Jill Bailey

Raintree

Chicago, Illinois

© 2004 Raintree
Published by Raintree, a division of Reed Elsevier, Inc.
Chicago, Illinois
Customer Service 888-363-4266
Visit our website at www.raintreelibrary.com

For information, address the publisher:
Raintree, 100 N. LaSalle, Suite 1200, Chicago, IL 60602

Project Editors: Marta Segal Block, Jennifer Mattson, Tamsin Osler
Production Manager: Brian Suderski
Designed by Ian Winton
Illustrated by Jim Chanell and Stuart Lafford

Planned and produced by Discovery Books

Library of Congress Cataloging-in-Publication Data:
Bailey, Jill.
Life in a garbage dump / Jill Bailey.
v. cm. -- (Microhabitats)
Includes bibliographical references and index.
Contents: The world of garbage -- Wonderful worms -- Creepy crawlies --
A free takeaway -- Smelly shelters -- Garbage dumps and humans.
ISBN 0-7398-6802-0 (lib. bdg.) -- ISBN 1-4109-0348-6 (pbk.)
1. Compost animals--Juvenile literature. 2. Waste disposal
sites--Juvenile literature. 3. Biodegradation--Juvenile literature. [1.
Waste disposal sites. 2. Compost animals. 3. Biodegradation.] I. Title.
II. Series.
QL110.5.B35 2004
577.5'5--dc21
2003002658

Printed and bound in the United States
08 07 06 05 04
10 9 8 7 6 5 4 3 2 1

Acknowledgments
The publishers would like to thank the following for permission to reproduce photographs:
*Front cover and p.25: Stephen Dalton/NHPA; p.6: John Forsdyke/Oxford Scientific Films; p.7: Heather Angel;
p.9: Eric Soder/NHPA; p.10: N.A. Callow/NHPA; p.12: Kjeli Sandved/Oxford Scientific Films; p.13: N.A. Callow/NHPA;
p.14: Kim Taylor/Bruce Coleman Collection; p.17: Martin Harvey/NHPA; p.18: Brian Kenney/Oxford Scientific Films; p.19:
Jane Burton/Bruce Coleman Collection; p.20: David Woodfall/NHPA; p.22: Breck P. Kent/Oxford Scientific Films;
p.23: Treat Davidson/FLPA; p.24: Michael Leach/NHPA; p.26: Minden Pictures/Oxford Scientific Films; p.27: Norbert
Rosing/FLPA; p.28: Silvestris Fotoservice/FLPA; p.29: T.C. Middleton/Oxford Scientific Films.*

Some words are shown in bold, **like this.** You can find out
what they mean by looking in the glossary.

Contents

The World of Garbage

For hundreds of millions of years—since long before the dinosaurs first appeared on Earth—leaves, fallen flowers and fruits, branches, and rotting tree trunks have littered the ground in wild places. Living **organisms** break down, or rot, these fallen remains until they crumble back into the soil as **compost,** a soft soil-like material that is rich in **nutrients.**

Garbage dumps are like giant heaps of compost that contain many different kinds of waste. Beneath their surface lies a rotting world where living creatures are hard at work breaking down garbage.

Gulls

Starling

Crow

Shrew

Cricket

Garbage that can be broken down by living organisms is called **biodegradable.** But many things in the garbage dump are not biodegradable. Cans and other metal objects, foam drinking cups, plastic, and packaging remain in the ground for hundreds of years without decaying.

Raccoon

Rat

Breakdown Specialists

Living things that break down dead plant and animal material—and even things like paper—are called **decomposers.** For some of them, a garbage dump is home, giving them a safe place to live and to raise their young. Other animals, the **scavengers,** visit the dump from time to time in search of food.

Masses of Microbes

The smallest **decomposers** are **bacteria,** exremely tiny specks of life. They are sometimes called **microbes,** because you can see them only with a microscope. Bacteria are around us everywhere. They arrive at the dump as **spores**—or bacteria in special hard skins—floating on air, stuck to the feet of flies, or on the garbage itself. Bacteria ooze juices that break down their food into a liquid pulp. The bacteria absorb this liquid food through the surface of their bodies.

Bacillus bacteria (shown here magnified under a microscope) are shaped like a rod. Other bacteria may be round or spiral in shape.

A Fungus Among Us

A **fungus** is an **organism** that feeds on decaying plant and animal material. More than one fungus are called fungi. Mushrooms and toadstools are fungi, and so are the molds that grow on old bread and fruit. The main fungus body is made of fine threads that dissolve food, then absorb it.

The threads of a fungus spread out over the rotting garbage in a dump. Slowly the garbage changes color and becomes softer until it crumbles into a powder. The fungi compete with the bacteria for the garbage. They produce chemicals called **antibiotics** to kill the bacteria.

Pinmold fungus is named for its hundreds of tiny, pinlike mushrooms.

See for Yourself

Put pieces of different kinds of fruit—apple, orange, and banana—in separate empty plastic containers in a garage, in a yard, or on a balcony. Over the next few weeks, observe what happens to them.

Add a spoonful of soil to the containers and see if that makes a difference.

Make sure you ask your parents' permission before you try this experiment.

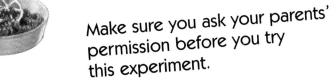

Wonderful Worms

Eating Without Teeth

Earthworms are very important **decomposers.** A worm has no teeth and it cannot chew, but it can eat rotting **vegetation,** as well as **bacteria, fungi,** and tiny creatures in soil and **compost.** A worm tunnels through dirt by sucking soil into its mouth. A pouch called the **gizzard,** located inside the **gut,** contains sand and grit to grind up the soil the worm eats. Digestive juices dissolve any plant and animal remains into liquid food that the worm can then absorb.

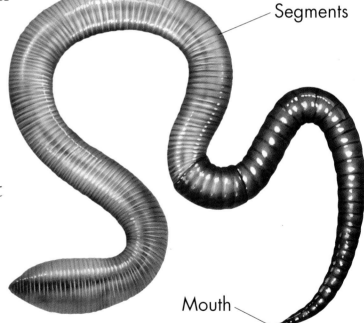

Segments

Mouth

An earthworm's body is divided up into segments. Each segment has four pairs of tiny bristles on the underside. All over the worm's body are glands that secrete a slimy fluid. This fluid helps the worm tunnel through the earth and prevents its skin from drying out.

Worm Casts

The leftover dirt passes out of the other end of the worm in squiggly-shaped piles called **worm casts.** Smaller creatures like bacteria can then feed on the casts, which are very rich in minerals.

The tunnels made by worms let air into the compost, allowing other decomposers living deeper down in the dump to breathe. Water and **nutrients** can trickle through the tunnels, too.

Worm casts are made of finely ground garbage and soil.

All Kinds of Worms

There are many different worms in a garbage dump. Brandling worms look like striped earthworms while potworms are like small white **maggots.** Some nematodes, or roundworms, are so tiny that a handful of rotting **compost** may contain several million of them. A clump of roundworms held under a microscope would look like human hair. These worms choose their food more carefully than earthworms. Some feed on **bacteria** and **fungi,** some eat rotting plants, and a few even eat other worms.

Potworms are less than half an inch (one centimeter) long, but here they are shown magnified.

Worm Farms

Worms can be used to make high-quality compost. Wormeries are special containers packed with dead **vegetation,** soil, and worms. The rotting material makes the wormery warm. As it gets warmer, the garbage rots even faster.

See for Yourself

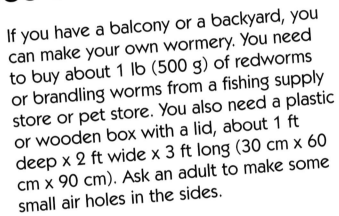

If you have a balcony or a backyard, you can make your own wormery. You need to buy about 1 lb (500 g) of redworms or brandling worms from a fishing supply store or pet store. You also need a plastic or wooden box with a lid, about 1 ft deep x 2 ft wide x 3 ft long (30 cm x 60 cm x 90 cm). Ask an adult to make some small air holes in the sides.

Shred several pounds of newspaper and sprinkle it with water to dampen it. Place the paper in the box. Make a hollow in the middle of it and add about 0.5 lb (250 g) of vegetable garbage. Do not add any meat—it gets smelly and will attract mice and rats.

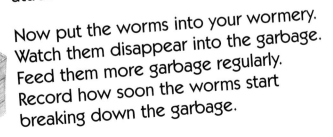

Now put the worms into your wormery. Watch them disappear into the garbage. Feed them more garbage regularly. Record how soon the worms start breaking down the garbage.

Garbage Creatures

Slow Movers

The slimiest animals in the garbage dump are the **gastropods**—the slugs and snails. They feed on freshly dumped and decaying plant matter. Some of the **nutrients** in their food are mixed back into the **compost.** Slugs and snails slide along with the help of muscles on their undersides. They produce a lot of slime to help them move smoothly. In winter a snail may seek out a sheltered corner of the dump, and then disappear inside its shell and seal up the opening. Then it goes to sleep.

Under a microscope, you can see the tiny teeth on the tongue of a snail.

Munching Maggots

Flies are common in garbage dumps. Fruit flies like to drink juices from decaying fruit. Dung flies come to hunt other flies. Bluebottles and houseflies visit the dump to lay eggs, and to feed, especially on rotting meat.

Houseflies lay up to 12 clusters of 100–150 eggs a year. The eggs hatch into **maggots,** which feed on the rotting garbage until they grow big and fat. Then the maggots form a tough skin around themselves and change into flies. A housefly egg can develop into a maggot and then into an adult fly in just two weeks.

Flies that visit garbage dumps may pick up **bacteria** and carry them away from the dump. If the flies land on food that people eat, the bacteria can cause stomachaches and more serious illnesses, too.

A bluebottle rests on an apple core.

See for Yourself

If you can find some snails, get a glass dish, and put some green leaves in one corner. Sprinkle water over the leaves and glass until they are really wet. Put your snails near the leaves. Watch the snails slide toward the leaves. Can you see the slime trails on the glass? When you have finished observing them, put the snails back where you found them.

Compost Critters

Animals have lived in **compost** for hundreds of millions of years. Yet some of them, such as millipedes, sow bugs, pill bugs, springtails, and cockroaches, still look very much like they did when they first appeared on earth. They feed on rotting garbage, **pollen** grains, and tiny pieces of fresh leaves, as well as on the droppings of other small creatures.

This cockroach is carrying an egg packet.

On warm days you may hear house crickets chirping in the compost heap. Crickets make their chirps by rubbing the left front wing, which has a row of pegs on its underside, against a ridge on the right front wing. Surprisingly, their ears are near their knees!

Garbage Pushers

With their flattened bodies, millipedes wriggle into narrow spaces, then hunch their backs to push the garbage apart, opening it up to the air. They rely on their awful smell to keep enemies away.

Sow bugs are related to crabs and lobsters. Their ancestors lived in the sea, and sow bugs still use gills for breathing just like fish and crabs. This is why they can live only in damp places.

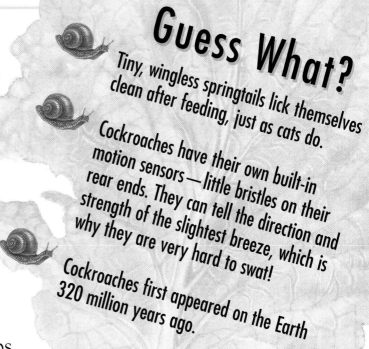

Guess What?

Tiny, wingless springtails lick themselves clean after feeding, just as cats do.

Cockroaches have their own built-in motion sensors—little bristles on their rear ends. They can tell the direction and strength of the slightest breeze, which is why they are very hard to swat!

Cockroaches first appeared on the Earth 320 million years ago.

Spot the Difference

Pill bugs, also known as roly-polies, are a type of sow bug. Both have armor-plated bodies. If in danger, however, pill bugs can roll up in a ball to protect their softer underparts. Sow bugs cannot do this.

Sow bug

Pill bug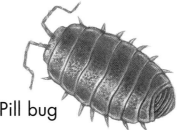

Pill bug curled up

Small but Deadly

The dump contains thousands of tiny killers. Ants, beetles, and centipedes hunt creatures even smaller than themselves. Unlike millipedes, which feed on plants, centipedes are meat-eaters. With their flat bodies, they can wiggle into the smallest cracks in search of prey. Pseudoscorpions look like tiny scorpions, but they have no tail. They feed on the same food as centipedes, using their large claws to grab flies, worms, mites, and other small animals. They then inject poison to paralyze their **prey.** Mites are some of the most common animals in **compost.** Like spiders, they have eight legs. Some mites feed on **bacteria** and **fungi,** but many hunt **grubs, maggots,** and springtails.

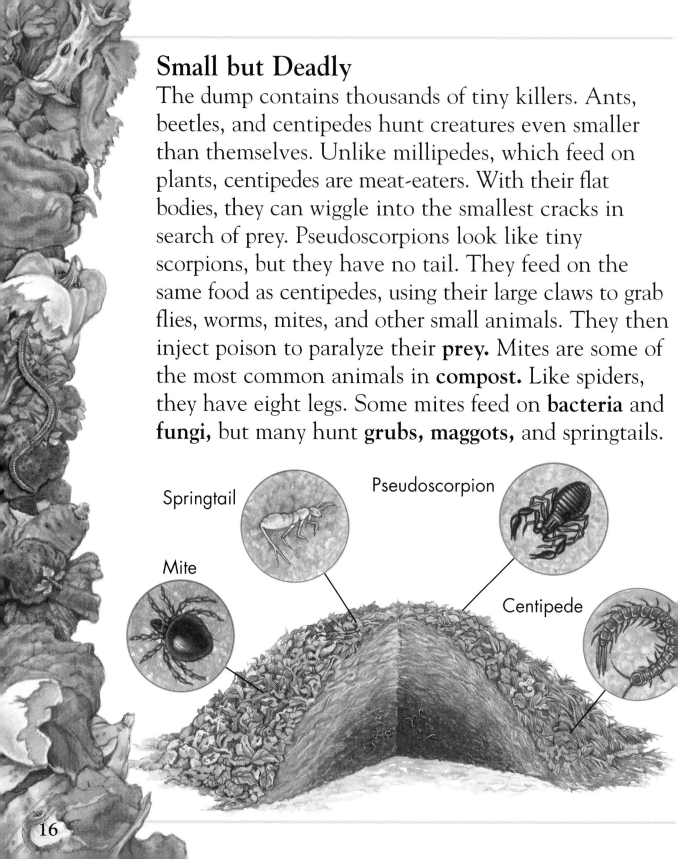

Springtail

Pseudoscorpion

Mite

Centipede

These dung beetles have collected a ball of dung. They will bury it, then the female will lay an egg in it. The grub will feed on the dung.

Nature's Gravediggers

There are many different kinds of beetles. Tiny feather-winged beetles feed on fungus **spores.** Other beetles eat paper and glue while many rove beetles and ground beetles are fierce hunters. Beetle eggs hatch into grubs and they, too, feast on rotting food or on very tiny compost animals. Burying beetles are gravediggers—they tunnel under the dead bodies of mice and other animals until the bodies sink into the ground. Then the beetles lay their eggs on the dead animal. Later, their grubs will feed on the flesh.

Easy Living

Living off Others

A garbage dump has a lot of food to offer visitors—both the rotting garbage itself as well as the animals that live in it. Frogs, toads, newts, salamanders, and shrews are **predators** that come to feed on **prey** such as insects and worms.

Frogs and toads feed on smaller creatures, but larger predators feed on them.

Spiders roam in dumps, too. Hunting spiders, including some wolf spiders, live in **burrows** in the garbage by day, and come out at night. They have eight eyes, and can see well enough at night to stalk and pounce on springtails, beetles, flies, and other small insects. Other kinds of spiders spin webbed traps across gaps in the garbage.

Hunters of the Night

At night the larger predators move in. Foxes look out for mice, while snakes will hunt frogs, toads, and even worms. Cats move silently, searching for mice and rats. A big garbage dump may support a large number of cats that have become wild. Skunks, raccoons, and foxes also come to pick over the trash, looking for not-too-rotten fruit, meats, and other tidbits.

Foxes explore the dump at night, looking through scraps and hunting for mice.

Guess What?

Toads do not use their tongues to push food down their throats. As they eat, they squeeze their large eyes down into their heads to help them swallow.

Newborn wolf spiders would easily get lost in garbage, but they climb up their mother's legs and ride on her back.

Old cans that still have a few scraps of food inside make good hunting grounds for little spiders. They spin their webs across the entrance to catch insects attracted by the smell of food.

19

Rest Stop

Garbage dumps are important feeding places for birds such as bald eagles, snowy owls, and, in Europe and the Middle East, storks. Crows, pigeons, and local garden birds also come to find food and to fight over the best bits. They will eat pieces of bread, cookies, and other human food, as well as any worm or **grub** that happens to show itself. As the sun sets, owls fly in to hunt for mice and shrews.

Attracted by a constant supply of food in garbage dumps, seagulls have adapted to life in places far away from water, feeding on scraps instead of fish.

Health Warning

Dumps can be dangerous for both birds and humans. Birds may choke on bits of plastic, or pick up diseases from the rotting food. **Bacteria** thrive in dumps, and birds like seagulls carry them from one place to another on their feet and bills. These bacteria can make humans as well as other birds sick.

See for Yourself

A **compost** heap in your backyard helps to attract songbirds such as robins, starlings, and finches. They come to feed on the earthworms, snails, insects, and bugs in the compost. In spring, they visit to look for nesting material.

Look out for a robin with its head cocked to one side. They cock their heads like that when they listen for worms underground.

Many birds have sharp, narrow beaks that help them pick up insects. Seed-eating birds usually have thicker, stronger beaks.

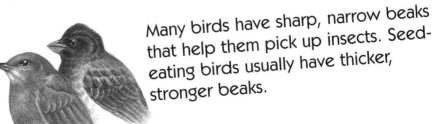

21

A Warm Shelter

As things rot, they give off heat. This means that garbage dumps and **compost** heaps are good places for some creatures to spend the winter. Garter snakes **hibernate** in large holes in or near the dump, sleeping through the cold months. In spring they soon fatten up on a diet of worms, slugs, crickets, toads, and other small creatures. Lizards, salamanders, and turtles also hibernate in dumps and compost heaps.

These baby garter snakes are warming themselves in the sun before moving off to hunt.

Egg Warmers

Because of their warmth, garbage dumps also make good **incubators** for eggs. Blacksnakes and garter snakes often lay their eggs in them. Snakes are cold-blooded, so they cannot warm their eggs with their own bodies like birds can. The American alligator and the Nile crocodile both build large piles of rotting **vegetation** in which to rear their eggs.

Alligator eggs develop in the warmth of a pile of rotting plants carefully built by their mother.

Guess What?

Deep down in a big dump the heat cannot escape and it gets very hot. Some garbage **bacteria** can live in temperatures of 212 °F (100 °C).

Whether a baby alligator is male or female depends on the temperature at which the eggs are kept. Eggs near the top of the nest mound are cooler than those at the bottom, and they hatch into baby male alligators, while the eggs at the bottom hatch into females.

Raiders of the Dump

Roving Rodents

Mice have lived close to humans ever since people first started growing wheat about 10,000 years ago. They dig their **burrows** in older, more stable parts of the dump. Rats and mice will eat almost everything humans eat, from grain and fruit to cookies, cheese, and peanut butter—even chocolate. They also gather and chew up pieces of paper, cardboard, straw, and cloth to make their nests. Rats also eat animal food such as eggs and young birds.

Rats and mice can multiply very rapidly when there is a good supply of food.

Disease Carriers

Rats can carry serious illnesses, like salmonella food poisoning and hantavirus—a potentially fatal disease that affects breathing in humans. To keep rats from coming close to places where humans live, it is important to control where and how garbage is dumped.

For rats, a garbage dump is a tasty feast.

25

Trouble with Bears

Bears find all sorts of things to eat in garbage dumps. They have a very good sense of smell, and can sniff out garbage from far away. In Churchill, Canada, on the coast of the Hudson Bay, polar bears gather every fall to wait for the winter ice to form so they can cross it to hunt seals. The long wait makes them hungry, so they raid garbage dumps and houses.

A polar bear searches through trash in Churchill, Canada.

Tourists come to see the bears, but they can be very dangerous. An angry bear could easily kill a human. Scientists shoot darts with **tranquilizers** into the bears, then fly the sleeping bears hundreds of miles away. Bears that keep coming back may end up in zoos.

Living Around the Garbage Dump

In many countries all around the world, poor and homeless people may live in garbage dumps, too. They make a living by picking over and sorting the garbage and then selling anything that could be of value—paper, tin, wood, scraps of leather and cloth, rubber tires, even plastic bags. The dumps are very unhealthy places to live, but sadly, very poor people sometimes have no other choice.

This child lives on a garbage dump in Madagascar.

Recycling Garbage

Much of the garbage created by humans remains in the dump for years and years, **polluting** the landscape. Finding new places to put trash is a growing problem in many countries. People need to reduce the amount of garbage they throw away, and reuse and **recycle** as much as possible. Scientists are trying to create special **bacteria** that will help plastic break down, so that it will take up less space in the dump. We can recycle cans for aluminum, bottles for glass, and old tires for rubber. Most paper can be recycled into new paper. And we can reuse metals and chemicals from old cars, TV sets, computers, and other household objects.

By **composting** as much as we can of our waste—including some paper and cardboard, grass cuttings, and wood shavings—we reduce the amount of garbage we send to the dump each week.

This garbage dump has been covered with soil and grass, and is now part of a nature reserve and park. Water draining off the old dump helps to fill the lake, where many ducks and other birds now live.

Reusing Dumps

When garbage dumps become full, they are covered with soil. In time, grass, plants, and flowers will begin growing there. In many cities, new parks, golf courses, or nature reserves have been created where there was once a dump. These places are now home to far more wildlife than ever lived in the rotting world of the garbage dump.

Guess What?

The garbage collected in the United States in just one day fills 100,000 garbage trucks.

Ancient garbage dumps are much like museums. People who study the history of human culture can discover details about everyday life in past times by studying what was thrown away.

In southern Israel, a garbage dump visited by migrating birds has been turned into a nature reserve. Now it attracts thousands of birds who rest there on their travels.

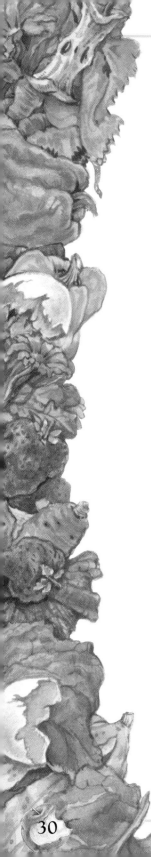

Glossary

antibiotic chemical that fights bacteria

bacterium (more than one are called bacteria) a kind of microbe that can be found almost anywhere

biodegradable term used to describe anything that can be decomposed by bacteria and other organisms

burrow to dig a hole or tunnel in the ground. A hole or tunnel used as an animal's home is called a burrow.

compost mixture of decaying plant and vegetable material. Compost can be added to soil to give the soil extra nutrients for growing things. The act of saving and combining plant and vegetable material for this purpose is called composting.

decomposer any organism that breaks down dead plant and animal material

fungus (more than one are called fungi) simple organism whose body is made from many tiny threads that spread over food, turn it into liquid, then absorb it

gastropod member of a group of animals with a soft body and no inside skeleton, such as a snail or slug

gizzard pouch located in the gut of some animals, such as birds and worms, that contains grit or sand to help grind up tough foods

grub young insect during the early stage when it looks nothing like its parents. Grubs look much like worms.

gut long tube that runs through an animal's body, through which food moves and wastes leave; intestine

hibernate to spend the winter in a special kind of deep sleep

incubator warm place where baby animals can develop inside their eggs until they are ready to hatch

maggot young fly during its wormlike grub stage

microbe any organism too small to see without a microscope

nutrient anything contained in food that helps keep a living thing healthy and active, such as vitamins or protein

organism any living thing

pollen powdery grains made by flowers to fertilize other flowers

polluting making something dirty or spoiled

predator animal that hunts another animal for food

prey animal that is hunted by another animal for food

recycle to process glass, paper, cans, and other materials in order to reuse them

scavenger animal that feeds on things that people or other animals have thrown away, or on dead remains of animals

spore small, hard case containing a bacterium that is carried from place to place by the wind. Some other kinds of organisms, such as fungi, produce spores containing everything needed to produce offspring.

tranquilizers drugs that cause sleepiness, often used to calm down animals

vegetation any form of plant life

worm cast little pile of fine soil that passes out of the end of a worm as it feeds

Further Reading

Greenaway, Theresa. *Minipets: Beetles*. Chicago: Raintree, 2000.

Greenaway, Theresa. *Minipets: Centipedes and Millipedes*. Chicago: Raintree, 2000.

Greenaway, Theresa. *Minipets: Worms*. Chicago: Raintree, 2000.

McGinty, Alice B. *Decomposers in the Food Chain*. New York: Rosen Publishing Group, 2002.

Roystone, Angela. *Recycling*. Chicago: Raintree, 1999.

Index